ISBN 978-0-331-75755-2
PIBN 11057379

1 MONTH OF
FREE
READING

at

www.ForgottenBooks.com

By purchasing this book you are eligible for one month membership to ForgottenBooks.com, giving you unlimited access to our entire collection of over 1,000,000 titles via our web site and mobile apps.

To claim your free month visit:
www.forgottenbooks.com/free1057379

English
Français
Deutsche
Italiano
Español
Português

www.forgottenbooks.com

Mythology Photography **Fiction**
Fishing Christianity **Art** Cooking
Essays Buddhism Freemasonry
Medicine **Biology** Music **Ancient**
Egypt Evolution Carpentry Physics
Dance Geology **Mathematics** Fitness
Shakespeare **Folklore** Yoga Marketing
Confidence Immortality Biographies
Poetry **Psychology** Witchcraft
Electronics Chemistry History **Law**
Accounting **Philosophy** Anthropology
Alchemy Drama Quantum Mechanics
Atheism Sexual Health **Ancient History**
Entrepreneurship Languages Sport
Paleontology Needlework Islam
Metaphysics Investment Archaeology
Parenting Statistics Criminology
Motivational

List of publications by members of the
Department of geology of the Leland
Stanford junior university.   1903.

LIST OF PUBLICATIONS

BY MEMBERS OF THE

DEPARTMENT OF GEOLOGY

OF THE

LELAND STANFORD JUNIOR UNIVERSITY,

———

JOHN CASPER BRANNER
JAMES PERRIN SMITH
JOHN FLESHER NEWSOM
RALPH ARNOLD

——— ———

STANFORD UNIVERSITY
CALIFORNIA
1903

143383

This list includes, besides the papers of the professors, the publications of students that have been prepared under the direction of the department.

Some of the papers in this list are available for exchanges. They may be obtained by addressing the authors at *Stanford University, California.*

PUBLICATIONS OF

JOHN CASPER BRANNER, Ph. D., LL. D.

PROFESSOR OF GEOLOGY

1884

1. The course and growth of the fibro-vascular bundles in palms. *Proceedings of the American Philosophical Society*, April 18, 1884, vol. XXI, pp. 459–483, 12 figs.

2. The pororóca or bore of the Amazon. *Science*, Nov. 28, 1884, vol. IV, pp. 488–492. Published as separate, with additional notes, 4 figs. Boston, 1885.

3. Rock inscriptions in Brazil. *American Naturalist*, Dec. 1884, vol. XVIII, pp. 1187–1192, 2 figs., 3 plates. The separates contain also pp. 1192a and 1192b.

4. Preliminary report of observations upon insects injurious to cotton, orange, and sugar-cane in Brazil. U. S. Department of Agriculture, Division of Entomology, Bulletin No. 4, pp. 63–69. Washington, 1884. The same report reprinted as a separate, Boston, 1884.

5. The Batrachichthys. *Science*, March 28, 1884, vol. III, p. 376, 1 fig.

6. Flexible sandstone. *American Naturalist*, Sept., 1884, vol. XVIII, p. 927.

1885

7. Inscripções em rochedos do Brazil. Translated by Dr. João Baptista Regueira Costa and published by the *Instituto Archeologico e Geographico Pernambucano*. 4 plates. Pernambuco, Brazil, 1885.

8. Glaciation of the Lackawanna valley. *Proceedings of the American Association for the Advancement of Science*, August, 1885, vol. XXXIV, pp. 212–214. Abstract, *Science*, 1885, vol. VI, pp. 221–222.

9. The reputation of the lantern-fly. *American Naturalist*, Sept., 1885, vol. XIX, pp. 835–838, 1 fig.

10. A Gitiranabóia. *Liberal Mineiro*, Ouro Preto, Brazil, Dec. 19, 1885.

11. Cotton in the Empire of Brazil ; the antiquity, methods, and extent of its cultivation, together with statistics of exportation and home consumption. Department of Agriculture, Special Report No. 8, pp. 79. Washington, 1885.

12. Cotton caterpillars in Brazil. Appendix V, pp. 49–54, of the Fourth Report of the U. S. Entomological Commission . . . on cotton worm and boll worm. Washington, 1885.

1886

13. Glaciation of the Wyoming and Lackawanna valleys. *Proceedings of the American Philosophical Society*, Feb. 19, 1886, vol. XXIII, pp. 337–357, 2 maps. Abstract, *Science*, 1886, vol. VIII, pp. 422.

14. Geographical and geological exploration in Brazil. *American Naturalist*, August, 1886, vol. XX, pp. 687–690.

15. Notes upon a native Brazilian language. *Proceedings of the American Association for the Advancement of Science*, August, 1886, vol. XXXV, pp. 329–330.

16. The thickness of the ice in Northeastern Pennsylvania during the glacial epoch. *American Journal of Science*, Nov., 1886, vol. CXXXII, pp. 362–366.

17. Geological map of Indiana, colored according to the scheme of the International Congress of Geologists, 2″ x 4″. Indianapolis, 1886.

1887

18. The railways of Brazil; reprinted from the *Railway Age*, July 8, 1887, vol. XII, pp. 470–473, with notes and additions, 26 pp., 2 maps. Chicago, 1887.

19. Annual report· of the geological survey of Arkansas for 1887, pp. 15. Little Rock, 1887.

20. Additional notes on the lantern-fly of Brazil. *Transactions of the New York Academy of Science*, Nov. 21, 1887, vol. VII, pp. 66–68.

21. Notes on the glacial striæ observed in the Lackawanna-Wyoming region. *Lackawanna Institute of History and Science*, vol. I, pp. 19–27. Scranton, 1887.

22. Topographical map in ten-foot contours of a portion of the Lackawanna valley between Scranton and Carbondale, Lackawanna county, in the Northern Anthracite coal field; scale 1600' = 1". Preliminary topographical map, Lackawanna valley sheets Nos. I and II. *Annual Report of the Second Geological Survey of Pennsylvania*, 1886. Harrisburg, 1887.

1888

23. The so-called gold and silver mines of Arkansas; an official report to Governor S. P. Hughes. *Arkansas Gazette*, Little Rock, Aug., 1888. *Engineering and Mining Journal*, New York, Aug. 18, 1888.

24. Notes on the fauna of the islands of Fernando de Noronha. *American Naturalist*, October, 1888, vol. XXII, pp. 861–871, 2 figs.

25. Notes on the Botocúdus and their ornaments. *Proceedings of the American Philosophical Society*, Nov. 16, 1888, vol. XXVI, pp. 171–173, 10 figs.

26. The Cretaceous and Tertiary geology of the Sergipe-Alagôas basin of Brazil. *Transactions of the American Philosophical Society*, 1888, vol. XVI, pp. 369–434, 5 plates, 10 figs., 4to.

27. On the manufacture of Portland cement. Chap. xxix of *Annual Report of the Geological Survey of Arkansas* for 1888, vol. II, pp. 291–302. Little Rock, 1888.

28. Introduction to "The Neozoic Geology of Southwestern Arkansas." *Annual Report of the Geological Survey of Arkansas* for 1888, vol. II, pp. xi–xiv. Little Rock, 1888.

29. Administrative report and introduction to "Geology of Western Central Arkansas." *Annual Report of the Geological Survey of Arkansas* for 1888, vol. I, pp. xv–xxxi. Little Rock, 1888.

30. Preface to "Geology of the Coal Regions." *Annual Report of the Geological Survey of Arkansas* for 1888, vol. III, pp. vii–x. Little Rock, 1888.

31. Glaciation; its relation to the Lackawanna–Wyoming valley. *Lackawanna Institute of History and Science*, vol. I, pp. 3–18, 4 plates. Scranton [Pa.], 1888.

32. Arkansas gold and silver mines; an official report to Governor S. P. Hughes in reply to certain charges. *Arkansas*

Democrat, Oct. 18, 1888.   *Engineering and Mining Journal,* Oct. 20, 1888, vol. XLVI, pp. 325–327.

1889

33. Arkansas State Weather Service. Appendix V of the Annual Report of the Chief Signal Officer, 1888, pp. 72–75. Washington, 1889.

34. A preliminary statement of the distribution of coal over the area examined by the Geological Survey (of Arkansas). *Arkansas Gazette,* Little Rock, Feb. 13, 1889.

35. The geology of Fernando de Noronha. *American Journal of Science,* Feb., 1889, vol. XXXVII, pp. 145–161, map, 7 figs.

36. The convict-island of Brazil — Fernando de Noronha. *Popular Science Monthly,* May, 1889, vol. XXXV, pp. 33–40.

37. The age and correlation of the Mesozoic rocks of the Sergipe-Alogôas basin of Brazil. *Proceedings of the American Association for the Advancement of Science,* 1889, vol. XXXVII, pp. 187–188.

38. The age of the crystalline rocks of Arkansas. *Proceedings of the American Association for the Advancement of Science,* 1889, vol. XXXVII, p. 188.

39. (With R. N. Brackett.) The peridotite of Pike County, Arkansas. *American Journal of Science,* 1889, vol. CXXXVIII, pp. 50–56, 1 fig, 1 plate. Reprinted in *Annual Report of the Geological Survey of Arkansas* for 1890, vol. II, pp. 378–391, 1 fig, 1 plate. Abstract, *Proceedings of the American Association for the Advancement of Science,* 1889, vol. XXXVII, pp. 188–189; *Neues Jahrbuch für Mineralogie,* 1893, pp. 500–501.

40. Analyses of Fort Smith clay shales. *Brick, Tile and Pottery Gazette,* June, 1889, vol. X, p. 114.

41. Bulding-stones of Arkansas. *Stone,* Oct., 1889, vol. II, pp. 92–93.

42. Geology of Arkansas. Abstract of a lecture delivered at Pine Bluff, Arkansas. Minutes of the *State Teachers' Association of Arkansas,* pp. 34–38. Little Rock, 1889.

1890

43. Some of the mineral resources of Northwestern Arkansas. *Arkansas Gazette,* Little Rock, Jan. 12, 1890; *Arkansas Press,* . Jan. 19, 1890.

44. Professor Hartt in Brazil. *Cornell Magazine*, Ithaca, N. Y., Feb., 1890, vol. II, pp. 186–192.

45. The training of a geologist. *American Geologist*, March, 1890, vol. V, pp. 147–160.

46. The æolian sandstone of Fernando de Noronha. *American Journal of Science*, April, 1890, vol. CXXXIX, pp. 247–257, 8 figs.

47. Geologia de Fernando de Noronha. No. 36 of the *Revista do Instituto Archeologico, e Geographico Pernambucano*. Pernambuco, Brazil, 1890, pp. 21, 1 map, 7 figs.

48. The relations of the state and national geological surveys to each other, and to the geologists of the country. *American Geologist*, Nov., 1890, vol. VI, pp. 295–309; *Science*, Aug. 29, 1890, vol. XVI, pp. 120–123; *Proceedings of the American Association for the Advancement of Science*, 1891, vol. XXXIX, pp. 219–237.

49. The pororóca, or bore, of the Amazon. *Popular Science Monthly*, Dec., 1890, vol. XXXVIII, pp. 208–215.

1891

50. A preliminary report upon the bauxite deposits of Arkansas, with locations and analyses. *Arkansas Gazette*, Little Rock, Jan. 8, 1891; *Arkansas Press*, Jan. 12, 1891; Biennial Report of the State Commissioner of Mines, Manufactures and Agriculture for 1893–94, pp. 119–126; Biennial Report of the same for 1895–96, pp. 105–112.

51. Bauxite in Arkansas. *American Geologist*, March, 1891, vol. VII, 181–183. *Science*, March 27, 1891, vol. XVII, p. 171. *Engineering and Mining Journal* (N. Y.), 1891, vol. LI, p. 114.

52. Introduction to "The Geology of Washington County" (Arkansas). *Annual Report of the Geological Survey of Arkansas* for 1888, vol. IV, pp. xi–xiv.

53. (With F. V. Coville.) A list of the plants of Arkansas. *Annual Report of the Geological Survey of Arkansas* for 1888, vol. IV, pp. 152–242. Little Rock, 1891.

54. Preface to "The Geology of Crowley's Ridge." *Annual Report of the Geological Survey of Arkansas* for 1889, vol. II, pp. xi–xix. Little Rock, 1891.

55. Preface to "Manganese : its Uses, Ores and Deposits." *Annual Report of the Geological Survey of Arkansas* for 1890, vol. I, pp. xxiii–xxvii. Little Rock, 1891.

56. Preface to "The Igneous Rocks of Arkansas." *Annual Report of the Geological Survey of Arkansas*, vol. II, pp. xi–xv. Little Rock, 1891.

57. Analyses of Hot Springs waters. Report of the Superintendent of the Hot Springs Reservation to the Secretary of the Interior, pp. 9–16. Washington, 1891.

58. David Starr Jordan, LL. D. (A biographical notice.) *The Delta Upsilon Quarterly*, New York, May, 1891, vol. IX, pp. 195–198.

1892

59. The mineral waters of Arkansas. *Annual Report of the Geological Survey of Arkansas* for 1891, vol. I, pp. 144, map. Little Rock, 1892.

60. The cotton industry in Brazil. *Popular Science Monthly*, 1892, vol. XL, pp. 666–674.

61. The training of a geologist. Third edition, 19 pp. San Francisco, 1892.

62. Preface to "Whetstones and the Novaculites of Arkansas." *Annual Report of the Geological Survey of Arkansas* for 1890, vol. III, pp. xv–xviii. Little Rock, 1892.

63. Preface to "The Iron Deposits of Arkansas." *Annual Report of the Geological Survey of Arkansas* for 1892, vol. I, p. xi. Little Rock, 1892.

1893

64. The lip and ear ornaments of the Botocúdus. *Popular Science Monthly*, Oct., 1893, vol. XLIII, pp. 753–757, 5 figs.

65. The supposed glaciation of Brazil. *Journal of Geology*, Chicago, vol. I, pp. 753–772, illustrated.

66. Preface to "Marbles and Other Limestones." *Annual Report of the Geological Survey of Arkansas* for 1890, vol. IV, pp. xvii–xxi. Little Rock, 1893.

67. Observations upon the erosion in the hydrographic basin of the Arkansas River above Little Rock. *Wilder Quarter-Century Book*, pp. 325–337. Ithaca, N. Y., 1893. Also separate, Ithaca, N. Y., 1893.

68. The coal fields of Arkansas. *Mineral Resources of the U. S.* for 1892, pp. 303–306, 1 fig. Washington, 1893.

69. Proverbs from the Portuguese. *The Overland Monthly* (San Francisco), May, 1893. Second series, vol. XXI, pp. 501–503.

1894

70. Elevations in the State of Arkansas. *Annual Report of the Geological Survey of Arkansas* for 1891, vol. II, pp. 77–152, 2 figs. Little Rock, 1894.

71. Observations upon the erosion in the hydrographic basin of the Arkansas River above Little Rock. *Annual Report of the Geological Survey of Arkansas* for 1891, vol. II, pp. 153–166.

72. Magnetic observations and meridian monuments established in Arkansas. *Annual Report of the Geological Survey of Arkansas* for 1891, vol. II, pp. 167–176, 10 figs. Little Rock, 1894.

73. Introduction to Sampson's "Preliminary List of the Mollusca of Arkansas." *Annual Report of the Geological Survey of Arkansas* for 1891, vol. II, pp. 179–180. Little Rock, 1894.

74. Introduction to Meek's "Catalogue of the Fishes of Arkansas." *Annual Report of the Geological Survey of Arkansas* for 1891, vol. II, pp. 216–220. Little Rock, 1894.

75. Bibliography of the geology of Arkansas. *Annual Report of the Geological Survey of Arkansas* for 1891, vol. II, pp. 319–340. Little Rock, 1894.

76. Introduction to and translation of the political constitutions of Brazil. *The Convention Manual of the Sixth New York State Constitutional Convention*, 1894. Part 2, vol. III, Constitution of the Empire, pp. 57–105. Constitution of the United States of Brazil, pp. 107–138. Albany, 1894.

77. Preface to "The Tertiary Geology of Southern Arkansas." *Annual Report of the Geological Survey of Arkansas* for 1892, vol. II, pp. xiii–xiv. Morrillton, 1894.

78. Report on road-making materials in Arkansas. U. S. Department of Agriculture, Office of Road Inquiry, Bulletin No. 4. Washington, 1894. Fourth biennial report of the Bureau of Mines, Manufactures and Agriculture (of Arkansas) for 1895–96, pp. 90–101. Little Rock, 1896. Also in fifth biennial report of that bureau for 1897–98, pp. 131–141.

79. The geological surveys of Arkansas. *Journal of Geology*, Chicago, vol. II, pp. 826–836.

80. The education of a naturalist. Commencement address at Leland Stanford Jr. University, May, 1894. *Daily Palo Alto*, May 30, 1894. Stanford University, 1894.

81. Os grés eólios de Fernando de Noronha. Instituto Archeologico e Geographico Pernambucano, 8 figs. Pernambuco, Brazil, 1894.

1895

82. (With John F. Newsom.) Syllabus of lectures on economic geology. Palo Alto, May, 1895, pp. 282.

83. (With John H. Means.) Great mountain railways. *The Chautauquan*, July, 1895, pp. 426–433.

84. Report upon the condition of the Geological Survey of Arkansas. Appendix to the Biennial Message of Governor Wm. M. Fishback to the General Assembly of the State of Arkansas, pp. 26–33. Morrillton, 1895.

1896

85. Our trade with South America. *The Argonaut*, San Francisco, Jan. 13, 1896.

86. Decomposition of rocks in Brazil. *Bulletin of the Geological Society of America*, 1895–96, vol. VII, pp. 255–314, 5 plates, 6 figs.

87. Thickness of the Paleozoic sediments in Arkansas. *American Journal of Science*, New Haven, Sept., 1896, vol. II, pp. 229–236, 8 figs.

88. The phosphate deposits of Arkansas. *Transactions of the American Institute of Mining Engineers*, vol. XXVI, pp. 580–598, map. New York, 1896.

89. Bibliography of clays and the ceramic arts. Bul. 143 of the U. S. Geological Survey, 114 pp., 2961 titles. Washington, 1896.

90. Review of "The Soil. By F. H. King." *Journal of Geology*, Chicago, vol. IV, p. 243.

91. (The decomposition of rocks in Brazil.) Editorial in *Journal of Geology*, Chicago, vol. IV, pp. 630–631.

92. (On the size of geologic publications.) Editorial in *Journal of Geology*, Chicago, vol. IV, pp. 214–217.

93. A supposta glaciação do Brazil. *Revista Brazileira*, April, 1896, vol. VI, pp. 49–55, 106–113. Rio de Janeiro, Brazil, 1896.

94. The study of science. (Part of a lecture delivered·at the Mount Tamalpais Military Academy.) *Overland Monthly*, San Francisco, Oct., 1896, Educational Department, pp. 26–30.

95. Abstract of F. Katzer's "Oldest Fossiliferous Beds of the Amazon Region." *Journal of Geology*, vol. IV, pp. 975–976. Chicago, 1896.

96. Review of the proceedings of the Indiana Academy of Sciences, geological subjects. *Journal of Geology*, vol. IV, p. 981. Chicago, 1896.

97. Geological map of Arkansas. Scale 80 ms. = 1".

1897

98. Note on "O fim da creação, pelo Visconde do Rio Grande." *Revista Brazileira*, Aug., 1897, pp. 254–255, Rio de Janeiro, Brazil ; also in *Annuario do Rio Grande do Sul* parar o anno de 1898, pp. 261–265. Porto Alegre [Brazil], 1897.

99. Bacteria and the decomposition of rocks. *American Journal of Science*, 1897, vol. CLIII, pp. 438–442, and as separate; abstract *Neues Jahrbuch für Mineralogie*, 1899, vol. II, Referate, 84.

100. The cement-materials of southwest Arkansas. *Transactions of the American Institute of Mining Engineers*, 1897, vol. XXVII, pp. 42–63, 6 figs. Also separate, 22 pp., map, and illustrations.

101. Reply to criticisms of R. T. Hill on "The Cement-materials of Southwest Arkansas." *Transactions of the American Institute of Mining Engineers*, 1897, vol. XXVII, pp. 945–946.

102. The bauxite deposits of Arkansas. *Journal of Geology*, April–May, 1897, vol. V, pp. 263–289, 2 plates, 2 figs. Also as separate with 10 pp. additional matter, Chicago, 1897.

103. (With J. F. Newsom.) The Red River and Clinton monoclines. *American Geologist*, July, 1897, vol. XX, pp. 1–13, 1 map and 3 figs. ; and separate.

104. Protection for American colleges. *The Nation*, New York, May 27, 1897, p. 395.

105. The introduction of new terms in geology. *Science*, June 11, 1897, vol. V, pp. 912–913; *Science*, July 23, 1897, vol. VI, pp. 133–134.

106. Mineral wealth of Arkansas. *Engineering and Mining Journal*, Aug. 7, 1897, p. 153.

107. Geology in its relations to topography. *Proceedings of the American Society of Civil Engineers*, Oct., 1897, vol. XXIII, No. 8, pp. 473–495; 1 plate, 16 figs.

108. Introduction to Ashley's "Geology of the Paleozoic Area of Arkansas South of the Novaculite Region." *Proceedings of the American Philosophical Society*, 1897, vol. XXXVI, pp. 217–220.

109. The former extension of the Appalachians across Mississippi, Louisiana, and Texas. *American Journal of Science*, Nov., 1897, vol. CLIV, pp. 357–371, 2 figs.   Abstract in *Report of the British Association for the Advancement of Science*, Toronto meeting, 1897, pp. 643–644; *Annales de Géographie*, 7me Année, Sept. 15, 1898, pp. 245–246; *Nature*, Nov. 18, 1897, vol. LVII, p. 70; *Journal of Geology*, Oct.–Nov., 1897, vol. V, pp. 759–760.

110. On the reporting of values to land owners by the State Geologist of Arkansas. *Hot Springs News*, July 12, 1897.

111. Review of "The Bedford Oölitic Limestone of Indiana. By T. C. Hopkins and C. E. Siebenthal, in 21st Ann. Rep. State Geologist of Indiana." *Journal of Geology*, July–Aug., 1897, vol. V, pp. 529–531.

112. The lost coal report of the Arkansas survey. Letter of Aug. 21, 1897. *Batesville Guard*, Sept. 3, 1897.

113. Review of the "Unpublished Reports of the Commissão Geologica do Brazil," published in the Boletim do Museu Paraense. *Journal of Geology*, Oct.–Nov., 1897, vol. V, pp. 756–757.

114. Review of Katzer's "Devonian fauna of the Rio Maecurú," published in the Boletim do Museu Paraense. *Journal of Geology*, vol. V, pp. 757–758. Chicago, 1897.

1898

115. Geology in its relations to topography (with discussion). *Proceedings of the American Society of Civil Engineers*, June, 1898, vol. XXXIX, pp. 53–95, 2 plates, 16 figs.

116. (With O. A. Derby.) On the origin of certain siliceous rocks. *Journal of Geology*, May–June, 1898, vol. VI, pp. 366–371. Abstract, *Neues Jahrbuch für Mineralogie*, 1900, vol. I, p. 408.

117. A geologist's impression (of the Grand Canyon of the Colorado, and Black Crater, Flagstaff, Arizona). *Land of Sunshine Magazine*, Aug., 1898, vol. IX, pp. 149–152, illustrated.

118. The Spanish University of Salamanca. *San Francisco Chronicle*, July 17, 1898, p. 12, illustrated. *Maryville College Monthly* for 1898, Maryville, Tenn.

119. Syllabus of elementary geology. 300 pp., 18 plates and 51 figs. Stanford University, 1898.

120. Review of "Earth Sculpture, by James Geikie. The Science Series, G. P. Putnam's Sons, New York, 1898." *Science*, Dec. 30, new ser., vol. VIII, pp. 957–959.

1899

121. Some old French place names in the State of Arkansas. *Modern Language Notes*, Feb., 1899, vol. XIV, No. 2, pp. 65–80. (Johns Hopkins University, Baltimore, Md.)

122. Review of "Volcanoes (*Science* series), by T. G. Bonney." *San Francisco Chronicle*, March 19, 1899.

123. The recent ascent of Itambé. *National Geographic Magazine*, vol. X, p. 183. Washington, 1899.

124. Notes upon the São Paulo sheet of the Commissão Geographica e Geologica de São Paulo, published in the *Revista Brazileira*, Rio de Janeiro, 1899, vol. ——, pp. ——; republished in the *Cidade de Santos*, Santos, Brazil, Jan. 10, 1900.

125. The São Paulo sheet of the topographic survey of São Paulo, Brazil. *Journal of Geology* (editorial), vol. VII, pp. 788–789. Chicago, 1899.

126. The maganese deposits of Bahia and Minas, Brazil. *Transactions of the American Institute of Mining Engineers*, Sept., 1899, vol. XXIX, pp. 756–770, 5 figs.

127. (With C. E. Gilman.) The stone reef at the mouth of Rio Grande do Norte. *American Geologist*, Dec., 1899, vol. XXIV, pp. 342–344, 2 figs.

128. A recife de pedra na foz do Rio Grande do Norte. Por J. C. Branner e C. E. Gilman. Traduzido por Dr. Alfredo de Carvalho. *Revista do Rio Grande do Norte*, 1900, Nos. 1, 2, Natal, Jan. e Fev., 1900, pp. 267–271.

129. Note upon "The Upper Silurian fauna of the Rio Trombetas, State of Pará, Brazil, and Devonian mollusca of the State of

Pará, Brazil.   By John M. Clarke.''   Archivos do Museu Nacional, vol. X, pp. 1–48, 49–174. *Journal of Geology*, Chicago, Nov.–Dec., 1899, vol. VII, pp. 813–814.

1900

130. Gold in Brazil. *Mineral Industry* for 1899, vol. VIII, p. 281.   New York, 1900.

131. Diamonds in Brazil. *Mineral Industry* for 1899, vol. VIII, pp. 221–222.   New York, 1900.

132. Ants as geologic agents in the tropics. *Journal of Geology*, Chicago, Feb.–Mar., 1900, vol. VIII, pp. 151–153, 3 figs.

133. The oil-bearing shales of the coast of Brazil. *Transactions of the American Institute of Mining Engineers*, Aug., 1900, vol. XXX, pp. 537–554.   Four half-tones, five line-drawings, one map of Bahia.   Review, *Neues Jahrbuch für Mineralogie*, 1901, vol. II, pp. 267–268.   Abstract, *Engineering and Mining Journal*, Sept. 15, 1900, vol. LXX, pp. 308–309.

134. (With J. F. Newsom.) Syllabus of economic geology; second edition, 368 + viii pp., 141 figs.   Stanford University, 1900.   (March 15th.)

135. South America. Encyclopædia Brittanica. *London Times* supplement.   London, 1900.

136. Review of ''A Preliminary Report on the Geology of Louisiana.   By G. D. Harris and A. C. Veatch.'' *Journal of Geology*, Chicago, April–May, 1900, vol. VIII, pp. 277–279.

137. Two characteristic geologic sections on the northeast coast of Brazil. *Proceedings of the Washington Academy of Science*, Aug. 20, 1900, vol. II, pp. 185–201, 3 plates, 5 figs.

138. Beach cusps. *Journal of Geology*, Chicago, Sept.–Oct., 1900, vol. VIII, pp. 481–484, 3 figs.

139. The zinc and lead region of North Arkansas. *Annual Report of the Geological Survey of Arkansas*, vol. V, 395 + xiv pp., 38 page plates, 92 figures in the text, and geologic atlas of 7 sheets.   Little Rock, December, 1900.   Reviewed by C. R. Keyes, *Journal of Geology*, Chicago, vol. IX, pp. 634–636.

1901

140. Review of ''A record of the geology of Texas, etc.   By F. W. Simonds.'' *Journal of Geology*, Chicago, vol. IX, p. 91.

141. Review of "Géologie et minéralogie appliqueés. Par Henri Charpentier. Paris, 1900." *Journal of Geology*, Feb.–Mar., 1901, vol. IX, pp. 198–199.

142. Os recifes de grés do Rio Formoso (Brazil). *Revista do Instituto Archeologico e Geographico Pernambucano*, No. 54, pp. 131–136, illustrated. Pernambuco, 1901.

143. The origin of travertine falls. *Science*, Aug. 2, 1901, vol. XIV, pp. 184–185.

144. The zinc and lead deposits of North Arkansas. *Transactions of the American Institute of Mining Engineers*, 27 illustrations, 32 pages, vol. XXXI, pp. 572–603. Republished in *Lead and Zinc News* of St. Louis, Mo., vol. II, Nov. 4, 1901, pp. 4–6; Nov. 11, 1901, pp. 4–6; Nov. 18, 1901, pp. 4–6; Nov. 25, 1901, pp. 4–5. Republished in *Arkansas Democrat* (semi-weekly). Little Rock, Ark., Nov. 24, 1901; Dec. 8, 1901; Dec. 22, 1901; Dec. 29, 1901. Abstract: *Engineering and Mining Journal*, New York, Nov. 30, 1901, pp. 718–719, 1 fig.

145. Editorial upon giant ripples. *Journal of Geology*, Chicago, Sept.–Oct., 1901, voí. IX, pp. 535–536.

146. The phosphate rocks of North Arkansas. *Arkansas Democrat*, Little Rock, Ark., Nov. 3, 1901. *Harrison Times*, Jan. 18, 1902.

147. Apontamentos sobre a fauna das Ilhas de Fernando de Noronha. Publicação do Instituto Archeologico e Geographico Pernambucano. 14 pp., 2 figs.; 8vo. Pernambuco, 1901.

· 1902

148. Depressions and elevations of the southern archipelagoes of Chile. By Francisco Vidal Gormaz. From the *Revista Nueva* of Santiago de Chile, 1901. Translation and introduction by J. C. Branner. *Scottish Geographical Magazine*, Edinburgh, Scotland, January, 1902, vol. XVIII, pp. 14–24, 1 map. Edinburgh, 1902;

149. Notes upon the surface geology of Rio Grande do Sul, Brazil. By James E. Mills. Edited from his letters by J. C. Branner. *American Geologist*, Feb., 1902, vol. XXIX, pp. 126–127.

150. The occurrence of fossil remains of mammals in the interior of the States of Pernambuco and Alagôas, Brazil. *American Journal of Science*, Feb., 1902, vol. CLXIII, pp. 133–137; 1 map, 1 half-tone plate.

151. Geology of the northeast coast of Brazil. *Bulletin of the Geological Society of America*, Rochester, vol. XIII, pp. 41–98, 16 figs., 9 plates.

152. The palm trees of Brazil. *Popular Science Monthly*, New York, vol. LX, pp. 386–412, 25 figs.

153. Discussion of Eric Hedburg's paper on "The Missouri and Arkansas Zinc Region." *Transactions of the American Institute of Mining Engineers*, vol. XXXI, pp. 1013–1014.

154. (With J. F. Newsom.) The phosphate rocks of Arkansas. Bul. 74, Arkansas Agricultural Experiment Station, Professor R. L. Bennett, Director, pp. 59–123. Fayetteville, Ark., Sept., 1902. 23 figures in text; 15 analyses.

155. Review of "The Scenery of England and the Causes to which it is due. By the Right Hon. Lord Avebury. New York, The Macmillan Co., 1902." *San Francisco Chronicle*, April 6, 1902.

156. Review of the "History of Geology and Palæontology. By Karl von Zittel. London and New York, 1902." *San Francisco Chronicle*, May 11, 1902.

157. Review of "The Earth's Beginning. By Sir Robert Stawell Ball. Appleton & Co., 1902." *San Francisco Chronicle*, June 22, 1902.

158. Syllabus of a course of lectures on elementary geology. Second edition, 370 pp., 109 figs., 25 plates. Stanford University, 1902.

159. The Carnegie Institution. *Science*, New York, Oct. 3, 1902. New series, vol. XVI, pp. 527–528.

In Press.

160. Bibliography of the geology, mineralogy, and paleontology of Brazil. (1288 titles.) In press by the *Bibliotheca Nacional do Rio de Janeiro*, Brazil.

161. Geologia elementar. (An elementary treatise on geology for the use of Brazilian students. Published in Portuguese.) Laemmert e:Cia., Rio de Janeiro. 256 MS. pages, 264 figures in the text, and 15 half-tone plates.

162. Biographical notice of James E. Mills. *Bulletin of the Geological Society of America*, Rochester, 1903, vol. XIV.

PUBLICATIONS OF
JAMES PERRIN SMITH, Ph. D.
PROFESSOR OF MINERALOGY AND PALEONTOLOGY

1892

1. Review of "Second Annual Report of the Geological Survey of Texas. Austin, 1890." *Neues Jahrbuch für Min., Geol. und Pal.*, 1892, II, 283.

1893

2. Die Jurabildungen des Kahlberges bei Echte. Jahrbuch d. Königl. Preussichen geologischen Landesanstalt und Bergakademie zu Berlin für das Jahr 1891, pp. 1–73.

1894

3. Age of the auriferous slates of the Sierra Nevada. *Bull. Geol. Soc. Amer.*, V, 243–258.

4. The Arkansas Coal Measures in their relation to the Pacific Carboniferous province. *Journal of Geology*, II, 187–204.

5. The metamorphic series of Shasta County, California. *Journal of Geology*, II, 588–612.

1895

6. Mesozoic changes in the faunal geography of California. *Journal of Geology*, III, 369–384.

7. Geologic study of the migration of marine invertebrates. *Journal of Geology*, III, 481–495.

1896

8. Review of "Palaeontographica Italica, vol. I, 1895. Pisa, 1896." *Journal of Geology*, IV, 242–243.

9. Classification of marine Trias. *Journal of Geology*, IV, 385–398.

10. Marine fossils from the Coal Measures of Arkansas. *Proc. Amer. Philos. Soc.*, XXXV, 213–285. Philadelphia.

1897

11. Comparative study of palaeontogeny and phylogeny. *Journal of Geology*, V, 507–524.

12. The development of Glyphioceras and the phylogeny of the Glyphioceratidae. *Proc. Cal. Acad. Sci.*, III Ser. (Geology), I, 103–128.

1898

13. Relations of paleontology to mining. *Engineering Journal,* Stanford University, II, 28–30.

14. The development of Lytoceras and Phylloceras. *Proc. Cal. Acad. Sci.*, III Ser. (Geology), I, 127–160.

15. Evolution of fossil Cephalopoda. In "Foot-notes to Evolution." By David Starr Jordan. D. Appleton & Co., New York, 1898. 8vo. ix, 229–255.

16. Geographic relations of the Trias of California. *Journal of Geology*, VI, 776–786.

1899

17. Larval stages of Schloenbachia. *Journal of Morphology*, XVI, 1–32.

1900

18. The development and phylogeny of Placenticeras. *Proc. Cal. Acad. Sci.*, III Ser. (Geology), I, 181–240.

19. The biogenetic law from the standpoint of Paleontology. *Journal of Geology*, VIII, 413–425.

20. Fossil Cephalopods in the Timan. A review of "Die Cephalopoden des Domanik im südlichen Timan. E. Holzapfel, Mém. Comité Géol. (Russie), Tome XII, No. 3 (1899)." *American Naturalist*, XXXIV, 830–831.

21. Russian Carboniferous Cephalopods. A review of "Nantiloidea et Ammonoidea du calcaire carbonifère. M. Zwetaew, Mém. Comité Géol. (Russie), Tome VIII, No. 4 (1898)." *American Naturalist*, XXXIV, 831.

22. Principles of paleontologic correlation. *Journal of Geology*, VIII, 673–697.

1901

23. The larval coil of Baculites. *American Naturalist*, XXV, No. 409, p. 39–49.

24. Triassic fossils from Eastern Siberia. A review of "Versteinerungen aus den Trias-Ablagerungen des Süd-Ussuri Gebietes in der ostsibirischen Küstenprovinz. A. Bittner, Mém. Comité Géol.

(Russie), vol. VII, No. 4 (1899)." *American Naturalist*, XXXV, 330.

25. The Upper Paleozoic fauna of Russia. A review of "Die Fauna einiger oberpalaeozoischer Ablagerungen Russlands. I. Die Cephalopoden und Gastropoden. N. Jakolew, Mém. Comité Géol. (Russie), vol. XV, No. 3 (1899)." *American Naturalist*, XXXV, 330-331.

26. (With Stuart Weller.) Prodromites, a new ammonite genus from the Lower Carboniferous. *Journal of Geology*, IX, 255-266.

27. Jurassic fossils from Alaska. A review of "Jura-Fossilien aus Alaska. J. F. Pompeckj, Verhandl. Kaiserl. Russischen Min. Gesell. St. Petersburg, Ser. II, Bd. XXXVIII, 1900." *American Naturalist*, XXXV, 420-421.

28. The Permian of Armenia. A review of " Das jüngere Palaeozoicum aus der Araxes-Enge bei Djulfa. G. von Arthaber, Beitr. Pal. und Geol. Oesterreich-Ungarns und des Orients, Bd. XII, No. 4, 1900." *American Naturalist*, XXXV, 421.

29. A review of "Die Triadische Cephalopoden-Fauna der Schiechling-Höhe bei Hallstatt. C. Diener, Beitr. Pal. und Geol. Oesterreich-Ungarns und des Orients, Bd. XIII, 1900." *American Naturalist*, XXXV, 421.

30. A review of "On some additional or imperfectly understood fossils from the Cretaceous rocks, etc. J. F. Whiteaves, Geological Survey of Canada. Mesozoic Fossils, I, Part IV, 1900." *American Naturalist*, XXXV, 422.

31. A review of "Die Schichten mit Venus konkensis am Flusse Konka. N. Sokolow, Mém. Comité Géol. (Russie), vol. IX, No. 5, 1899." *American Naturalist*, XXXV, 422.

32. A review of "Die Gastropoden der Esinokalke, nebst einer Revision der Gastropoden des Marmolatakalke. E. Kittl, Annalen d. K. K. Naturhist. Hofmuseums (Wien), Bd. XIV, No. 1-2, 1899." *American Naturalist*, XXXV, 422.

33. The border-line between Paleozoic and Mesozoic in Western America. *Journal of Geology*, IX, 512-521.

1902

34. The Carboniferous Ammonoids of America. U. S. Geological Survey, Monograph XLII, 1-205, 29 plates, 4to. Washington, D. C.

35. Ueber Pelecypoden–Zonen in der Trias Nord Amerikas. *Centralblatt für Mineralogie, Geologie und Palæontologie*, Stuttgart. Bd. III, 689–695.

In Press

36. The comparative stratigraphy of the Trias of Western America. *Proc. California Acad. Sci.*, III Ser. (Geology), II.

37. (With Alpheus Hyatt.) The Triassic Cephalopod genera of America. U. S. Geological Survey, Professional Papers, No. —. About 200 p., 61 plates, 4to. Washington, D. C.

PUBLICATIONS OF

JOHN FLESHER NEWSOM, PH. D.

ASSOCIATE PROFESSOR OF MINING

1895

1. A relief map of Arkansas. *Proc. Ind. Acad. Sci.*, 56.

2. (With J. C. Branner.) Syllabus of lectures on economic geology. 282 p. Palo Alto, May.

1897

3. A geological section across Southern Indiana from Hanover to Vincennes. *Proc. Ind. Acad. Sci.*, 250–253.

4. The Knobstone group in the region of New Albany, Indiana. *Proc. Ind. Acad. Sci.*, 253–256.

5. (With J. C. Branner.) The Red River and Clinton monoclines, Arkansas. *American Geologist*, XX, 1–13, map, 3 figs. Minneapolis.

6. Geological relief map of Morrison's Cove and the adjacent region, Pennsylvania. Scale, 1 inch = 3500 ft. Bloomington, Ind.

7. Geological relief map of Allamakee County, Iowa. Scale, 1 inch = 1 mile. Bloomington, Ind.

8. Relief map of Crater Lake, Oregon. Scale, 1 inch = 4000 ft. Bloomington, Ind.

9. Sectional geological relief map of the Leadville region, Colorado. Scale, 1 inch = 2640 ft. Bloomington, Ind.

10. Geological relief map of a section across southern Indiana. Scale, 1 inch = 1 mile. Bloomington, Ind.

11. Relief map of an ideal restoration of the Marysville Buttes, California. Scale, 1 inch = 5208 ft. Bloomington, Ind.

1898

12. A geological section across Southern Indiana, from Hanover to Vincennes. *Journal of Geology*, Chicago, VI, 250–256. 1 plate (XI).

13. (With J. A. Price.) Notes on the distribution of the Knobstone group in Indiana. *Proc. Ind. Acad. Sci.*, 289–291, map.

1899

14. The effect of sea barriers upon ultimate drainage. *Journal of Geology*, Chicago, VII, 445–451, 4 figs.

1900

15. (With J. C. Branner.) Syllabus of economic geology, second edition, illustrated ; 368 + viii p.   Stanford University, March.

1902

16. Drainage of southern Indiana. *Journal of Geology*, Chicago, X, 166–181, map.

17. (With J.C. Branner.) The phosphate rocks of Arkansas. *Bul. No. 74 Ark. Agr. Exp. Sta.*, 1 plate, 23 figs.   Fayetteville, Ark.

18. A natural gas explosion near Waldron, Indiana. *Journal of Geology*, Chicago, X, 803–814, 5 figs.

In Press

19. A geologic and topographic section across southern Indiana, from Hanover to Vincennes, with a discussion of the Knobstone group in the State of Indiana. *Indiana Department of Geology and Natural History.*

20. Clastic dikes. *Geological Society of America*, XIV.

PUBLICATIONS OF
DORSEY ALFRED LYON, A. M.
INSTRUCTOR IN METALLURGY

1902

1. The Mount Baker mining district. *Wash. Geol. Surv.*, I, Ann. Rep. for 1901; Part II, 79. Olympia.

2. The State Creek mining district. *Wash. Geol. Surv.*, I, Ann. Rep. for 1901; Part II, 83. Olympia.

3. The Thunder Creek mining district. *Wash. Geol. Surv.*, I, Ann. Rep. for 1901; Part II, 89. Olympia.

4. The Darrington mining district. *Wash. Geol. Surv.*, I, Ann. Rep. for 1901; Part II, 99. Olympia.

PUBLICATIONS OF
RALPH ARNOLD, Ph. D.
ASSISTANT IN GEOLOGY

1902

1. (With Delos Arnold.) The Marine Pliocene and Pleistocene stratigraphy of the coast of Southern California. *Journal of Geology*, Chicago, X, 117–138, 2 maps, 7 figs., 5 plates.

2. (Fossils from Ponta de Pedras, Pernambuco, Brazil.) *Bul. Geol. Soc. Amer.*, Rochester, XIII, 47.

3. Bibliography of the literature referring to the geology of Washington. *Wash. Geol. Surv.*, I, Ann. Rep. for 1901 ; Part VI, 16. Olympia.

In Press

4. The paleontology and stratigraphy of the Marine Pliocene and Pleistocene of San Pedro. *Mem. Cal. Acad. Sci.*

PUBLICATIONS FROM THE GEOLOGICAL LABORATORY BY STUDENTS

Anderson, F. M.

Some Cretaceous beds of Rogue River Valley, Oregon. *Journal of Geology*, Chicago, 1895, III, 455–468, 5 figs.

Ashley, George H.

An illustration of the flexure of rock. *Proc. Cal. Acad. Sci.*, 2d Ser., III, 319–324, 3 figs. San Francisco, 1893.

Studies in the Neocene of California. *Journal of Geology*, III, 434–454, 1 map, 2 plates. Chicago, 1895.

The Neocene stratigraphy of the Santa Cruz Mountains of California. *Proc. Cal. Acad. Sci.*, V, 273–367, 1 fig. San Francisco, 1895.

Geology of the Paleozoic area of Arkansas south of the novaculite region. *Proc. Amer. Phil. Soc.*, XXXVI, 217–318, 2 maps, 2 plates, 37 figs. Philadelphia, 1897.

Barber, Wm. B.

(With E. H. Nutter.) On some glaucophane and associated schists in the Coast Ranges of California. *Journal of Geology*, X, 738–744. Chicago, 1902.

Drake, Noah Fields

The topography of California. *Journal of Geology*, V, 563–578, 1 plate. Chicago, 1897.

A geological reconnoissance of the coal fields of the Indian Territory. *Proc. Amer. Phil. Soc.*, XXXVI, 326–419, 9 plates, 4 figs. Philadelphia, 1897.

Relief map of the State of California. Scale, 1 inch = 12 miles. Stanford University, 1896.

Gilman, C. E.

(With J. C. Branner.) The stone reef at the mouth of Rio

Grande do Norte, Brazil. *American Geologist*, XXIV, 342–344, 2 figs. Minneapolis, 1899.

Gwartney, J. G.

Relief map of the Palo Alto, San Jose, and Mt. Hamilton sheets of the U. S. Geological Survey. Scale, 1 inch = .71 mile. Stanford University, 1897.

Hopkins, Thomas C.

Marbles and other limestones (of Arkansas). *Ann. Rep. Geol. Surv. Ark. for 1890*, IV, xv + 443, 28 plates, 14 figs. Little Rock, 1893.

(With F. W. Simonds.) The geology of Benton County (Arkansas). *Ann. Rep. Geol. Surv. Ark. for 1891*, II, 1–75, 1 map. Little Rock, 1894.

Means, John H.

Carboniferous limestones on the south side of the Boston Mountains. *Ann. Rep. Geol. Surv. Ark. for 1890*, IV, 126–136. Little Rock, 1893.

Nutter, Edward Hoit

Sketch of the geology of the Salinas Valley, California. *Journal of Geology*, IX, 330–336; 2 maps, 6 figs. Chicago, 1901.

(With Wm. B. Barber.) On some glaucophane and associated schists in the Coast Ranges of California. *Journal of Geology*, X, 738–744. Chicago, 1902.

Purdue, A. H.

Relief map of San Francisco Peninsula. Scale, 1 inch = 1666 feet. Stanford University, 1894.

Shedd, Solon

Relief map of Oregon. Scale, 1 inch = 12 miles. Stanford University, 1896.

Siebenthal, C. E.

The geology of Dallas County (Arkansas). *Ann. Rep. Geol. Surv. Ark. for 1891*, II, 277–318; 1 map, 1 fig. Little Rock, 1894.

CPSIA information can be obtained
at www.ICGtesting.com
Printed in the USA
LVHW050031271118
598300LV00027B/1077/P

9 780331 757552